# STATIC ELECTRICITY
## (WHERE DOES LIGHTNING COME FROM)

### 2nd Grade Science Workbook
### Children's Electricity Books Edition

**BABY PROFESSOR**
EDUCATION KIDS

Speedy Publishing LLC
40 E. Main St. #1156
Newark, DE 19711
www.speedypublishing.com

The study of lightning is known as fulminology.

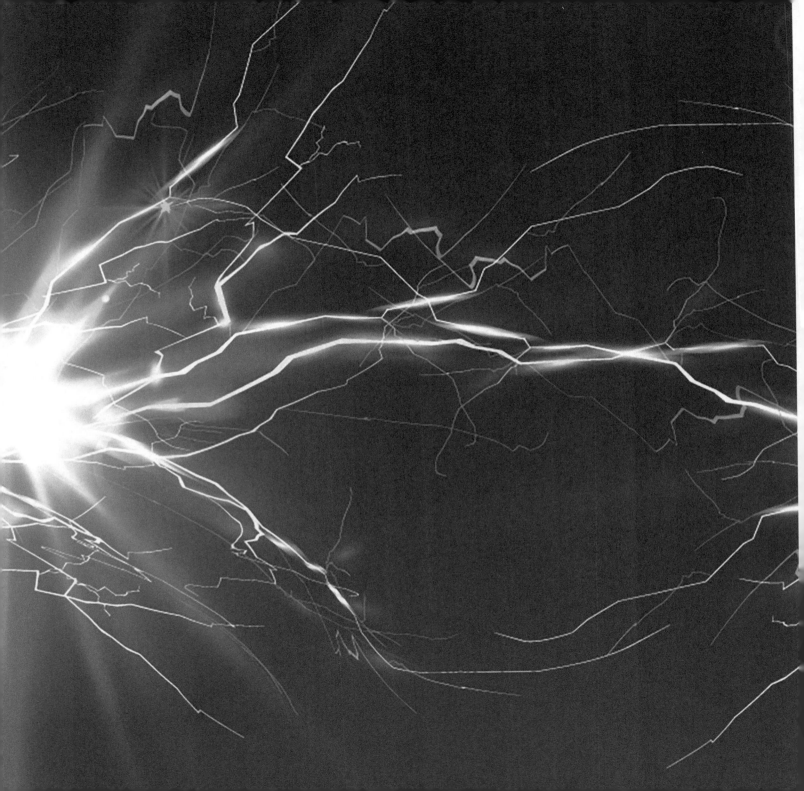

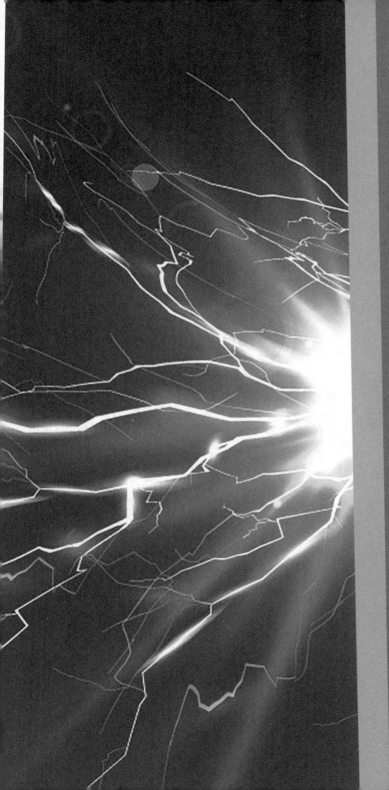

Static electricity is the build up of an electrical charge on the surface of an object.

It's called "static" because the charges remain in one area.

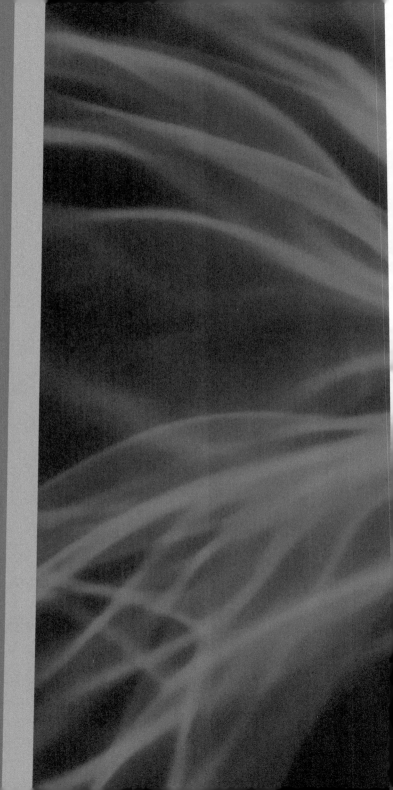

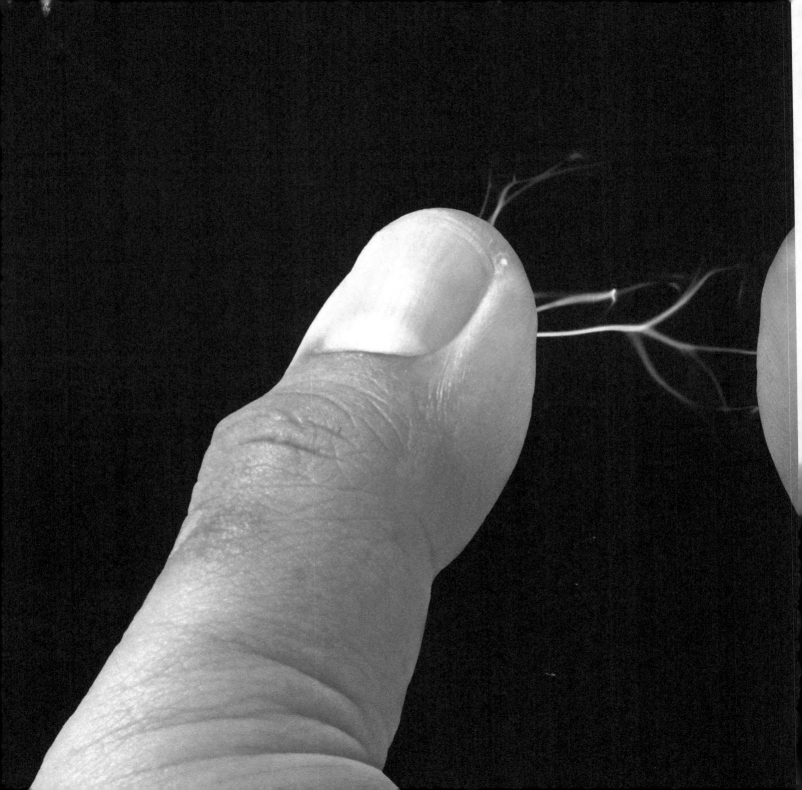

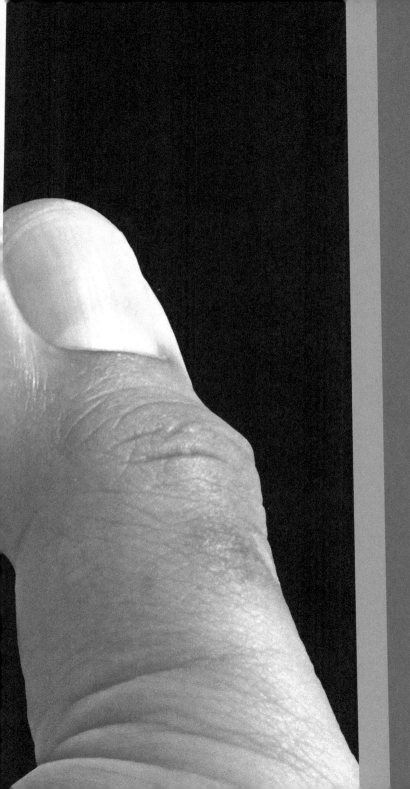

A static electric charge is created whenever two surfaces contact and the electrons move from one object to another.

A spark
of static
electricity
can measure
thousands
of volts.

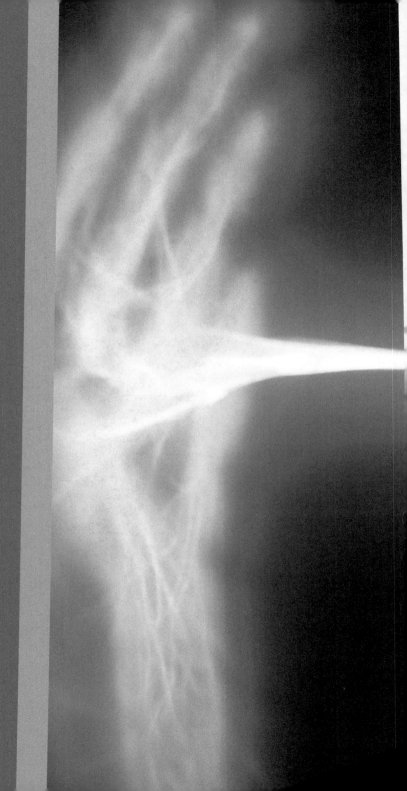

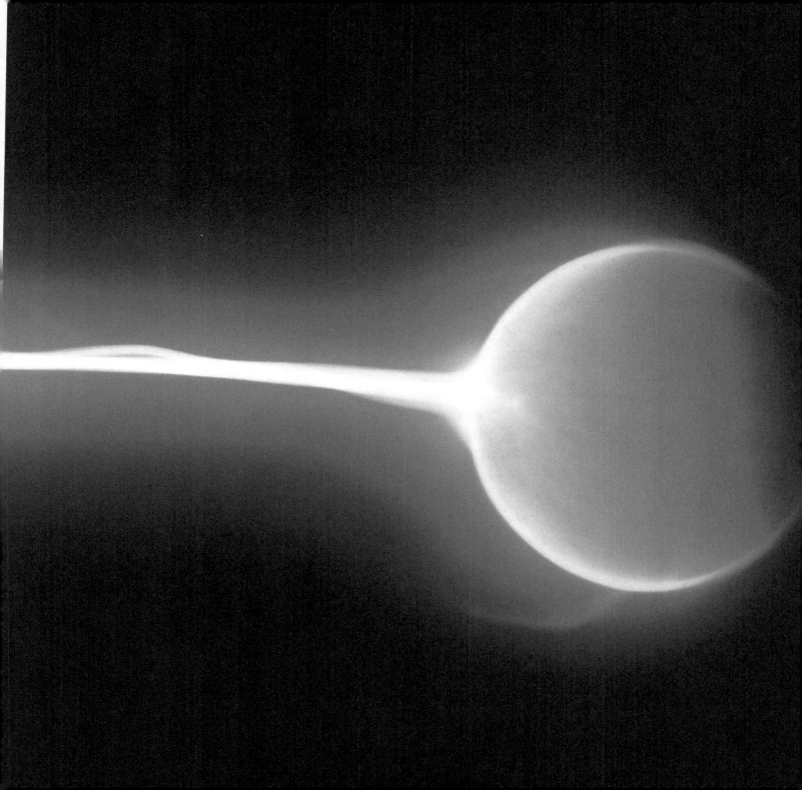

We see static electricity every day. It can even build up on us.

Lighting is also
a powerful
and dangerous
example
of static
electricity.

Lightning is a powerful burst of electricity that happens very quickly during a thunderstorm.

Lightning is caused by an electrical charge in the atmosphere that is unbalanced.

Within a thundercloud, many small bits of ice bump into each other as they move around in the air.

As they bump
into each other,
they create
an electric
charge.

When the charge connects with electrical charges on the ground, lightning strikes.

Lightning can occur inside the clouds, between the clouds and from clouds to the ground.

Lightning is usually produced by cumulonimbus clouds that are very tall and dense.

Lightning can have 100 million to 1 billion volts, and contains billions of watts.

Most lightning occurs over land rather than oceans, with around 70% of it occurring in the Tropics.

The temperature of a lightning flash is 15,000 to 60,000 degrees Fahrenheit.

On Earth, the lightning frequency is approximately 40-50 times a second.

A lightning
flash is no
more than one
inch wide.

Visit

**BABY PROFESSOR**
EDUCATION KIDS

# www.BabyProfessorBooks.com

to download Free Baby Professor eBooks
and view our catalog of new and exciting
Children's Books

CPSIA information can be obtained
at www.ICGtesting.com
Printed in the USA
LVHW061138200420
654120LV00015B/2725

Visit

**BABY PROFESSOR**
EDUCATION KIDS

# www.BabyProfessorBooks.com

to download Free Baby Professor eBooks
and view our catalog of new and exciting
Children's Books